Attendant
Operator
Chemical Plant

Manoj Dole

ISBN 978-93-5438-115-7
© Manoj Dole 2021
Published in India 2021 by Pencil

A brand of
One Point Six Technologies Pvt. Ltd.
123, Building J2, Shram Seva Premises,
Wadala Truck Terminal, Wadala (E)
Mumbai 400037, Maharashtra, INDIA
E connect@thepencilapp.com
W www.thepencilapp.com

Author biography

MANOJ DOLE is an Engineer from reputed University. He is currently working with Government Industrial Training- Institute as a lecturer from last 12 Years. His interest include- Engineering Training Material, Invention & Engineering Practical- Knowledge etc.

CONTENTS

Attendant Operator Chemical Plant

01] In case of bleeding, take treatment Of

D] cold 3" and rest

A] spray cold water

B] Bandage immediately -----.

B] Enquire about the accident thought treatment

02] in case of an accident, the victim should im

A] Asked to take rest

C] Attended immediately

D] leave him

03] First aid is given to an injured or ill person primarily....

A] Save life

B] Prevent further deterioration of the muff's

C] Give best possible comfort

D] All of these

04] Colour code for Bins for waste paper segregation is ----

A] blue Colour

B] Yellow Colour

C] Red Colour

D] Green Colour

05] In Japanese Seiko stands for --------------

A] Shine

B] Sort
C] Standardize
D] Sustain

06] Benefit of SS system is ------
A] Increase in productivity
B] Increase in quality
C] Reduction in wastage of time
<u>D] All of these</u>

03] SAFETY PRECAUTION 09
07] Safety is -----------
A] nobody's business
<u>B] every bodise business</u>
C] Some bodies business
D] The organization business

08] For basic categories of safety signs are available The meaning of"prohibition" sign ----
<u>A] shows it must not be done</u>
B] Shows what must be done
C] Warns the hazard or danger
D] Gives information of safety provision

09] Which one is a workshop safety?
<u>A] Keep shop floor clean and free from grease, oil or other slippery materials</u>
B] Stop the machine before changing the speed
C] Don't use cracked or chipped tools
D] Don't try to stop a running machine with hand

10] In Personal Protect Equipment (PPE] HELMET is

used to
A] protect head
B] Protect eyes
C] Protect hands
 D] Protect ears

11] Which of the following belongs to general safety?
A Have a worker in good attitude
 B] The work clean and clear
C] Concentrate on your work
D] Keep the floor and gangways clean and clear

12] While grinding, which is used to protect the eyes?
A] Dark green glass
B] Mask
C] Sun glasses
D] Safety goggles

13] Which of the following is done for machine safety?
A] Check the oil level before starting the machine
B] Do things in a methodical way
C] Keep the floor and gangways clean and clear
D] Don't use dies and scarves

14] In Personal Protect Equipment (PPE], 'sleeves' is used
to protect ----------
A] Face
B] Eyes
C] Ears
D] Hands

15] ABC stands for --------------

A] Automatic Breathing Control
B] Automatic Blood Control
C] Airway Breathing Circulation
D] Automatic Blood Circulation

04] Fire & FIRE EXTINGUISHERS

16] To put off"Class B" fire, the types of fire extinguisher used is ....
A] dry power
B] Carbon dioxide
C] Jet of water
D] Foam type

17] Which type of fire extinguisher is used to put off general fire?
A] Water type Extinguisher
B] Foam type Extinguisher
C] Dry chemical powder Extinguisher
 D] Carbon dioxide (C02] Extinguisher

18] One micrometer (U] is equal to...
A] 0.1mm
B] 0.01mm
C] 0.001mm
D] 0.0001mm

19] The caliper meant for measuring the width of a slot is...
A] Odd leg caliper
B] Outside caliper
C] Jenny caliper
D] Inside calliper

20] The size of the dividers are specified by the ----------
A] Total length of legs
B] Distance between the points when fully opened
C] Length of legs without points
D] distance between the pivot and the point

21] The instrument used to mark parallel lines, parallel to the datum edge is -
A] jenny caliper
B] Divider
C] Outside calliper
D] Inside calliper

22] Which one of the following is an indirect measuring tool?
A] Outside caliper
B] Vernier calliper
C] Steel rule
 D] Outside micrometer

23] For cutting thin tubing, the most suitable pitch of the hacksaw blade is...
A] 1.8mm
B] 1.4mm
C] 1mm
D] 0.8mm

24] For cutting solid brass, the most suitable pitch of the hacksaw blade is...
A] 1.8mm
B] 1.4mm

C] 1mm
D] 0.8mm

25] A new hacksaw blade after a few strokes becomes loose because of the...
A] <u>Stretching of the blade</u>
B] Wing-nut threads being worn out
C] Wrong pitch of the blade
D] Improper selection of the set of saws.

26] While cutting small diameter pipes, it is advisable to watch regularly and ensure that...
A] The cut is along the curved line
B] <u>More saw teeth are in contract</u>
C] The work is not overheated
D] Proper balancing of hacksaw is maintained

27] The vice clamps are used to...
A] Protect hard jaws
B] Clamp the work pieces rigidly
C] <u>Protect the finished surfaces</u>
D] Prevent the movable jaw being filed

28] The reference surface during marking is provided by the...
A] Surface gauge
B] Workpiece
C] Drawing of the work
D] <u>Marking table surface</u>

29] The size of an engineer's vice is specified by the...
A] Length of the movable jaw

B] <u>Width of the jaws</u>
C] Height of the vice
D] Maximum opening of the jaws

30] The part of the universal surface gauge which helps to draw a parallel line along a datum edge is the..
A] Rocker arm
B] Snug
C] Fine adjustment screw
D] <u>Guide pins</u>

31] Scribers are made of...
A] Mild steel
B] <u>High carbon steel</u>
C] Brass
D] Cast iron

32] Portion of the hammer used for fixing the handle is...
A] Face
B] Peen
C] Cheek
D] <u>Eye hole</u>

33] Weight of the hammer for the marking purpose is...
A] <u>250g</u>
B] 500g
C] 1 kg
D] 2 kgs

34] The size of the dividers are specified by the...
A] Total length of the legs
B] Distance between the points when fully opened

C] Length of legs without the points
D] <u>Distance between the pivot and the point</u>

35] The included angle of the groove of 'V' block is always....
A] 45°
B] 60°
C] 90°
D] <u>120°</u>

36] 'V' blocks are available in grades of...
A] <u>A & B</u>
B] A,B & C
C] 1,2 & 3
D] 1 & 2

37] 'V' blocks of grade 'B' are made of
A] <u>Cast iron</u>
B] Mild steel
C] Steel
D] Cast steel

38] Name the punch used to locate the centre.
A] Prick punch 30°
B] Prick punch 60°
C] <u>Centre punch</u>
D] Dot punch

39] The point angle of centre punch is --------
A] 30°
B] 50°
c] <u>900</u>

D] 1200

40] Punches are used for forming ---------of any shape
A] Holes
B] Mining
C] Knurling
D] Reaming

41] Generally the length of the handle of the vice is ---------
-
A] 1.5 times the normal size of the vice
B] 2.5 times the normal size of the vice
C] 3.5 times the normal size of the vice
D] 4.5 times the normal size of the vice

42] Bench vice spindle is made of ....
A] mild steel
B] Cast iron
C] Tool steel
D] Bronze

43] The convexity of files helps...
A] To file concave surfaces
B] To file convex surfaces
C] To prevent rounding of edges of work
D] The file to become straight when pressure is applied

44] Which file used for filling wood, leather and other soft
material? .
A] Single cut file
B] Double cut file
c] Rasp cut file

D] Curved cut file

45] File used is used for ------------
A] Cleaning the work piece
C] Renewing the file teeth
B] cleaning the file teeth
D] Cleaning the chips

46] File card is used to --------
A] Clean the work piece
C] Renew the file teeth
B] Clean the file teeth

47] The point angle of scriber is -----------
A] 30°
B] 60°
C] 5° to 10°
D] 12° to 15°

48] The least count of vernier calliper is (main scale = 49 division, vernier scale = 50 division]
A] 0.1 mm
B] 0.01 mm
C] 0.001 mm
D] 0.02 mm

49] The type of measurement made by using a Vernier Calliper is -------
A] Direct measurement
B] Indirect measurement
C] 90°] (a] 81 (b]
D] None of these

50] The least count of a vernier height gauge in the metric system is
A] 0.05 mm
B] 0.1 mm
C] <u>0.02 mm</u>
D] 0.001 mm

51] The least count of a vernier height gauge in the british system is
A] 0.05"
B] <u>0.001"</u>
C] 0.002"
D] 1"

52] For marking purposes, a vernier height gauge must be used on the
A] bed of a machine tool
B] <u>surface plate</u>
C] square block
D] any flat surface

53] The reading of a vernier height gauge is similar to that of a
A] <u>vernier caliper</u>
B] depth micrometer
C] dial test indicator
D] gauge

54] The part which slides on the beam of a vernier height gauge is known as a
A] base
B] beam scale

C] scriber
D] <u>vernier slide</u>

55] The size of a vernier height gauge is specified by the
A] height of the vernier scale
B] <u>height of the beam</u>
C] width of the beam
D] size of the base

56] The base of the vernier height gauge is generally made
out of
A] cast iron
B] <u>steel</u>
C] aluminium alloy
D] tungsten carbide

57] Accuracy or least count of a metric outside micrometer
is ---------
A] 0-1 mm
B] <u>0.01 mm</u>
C] 0.001 mm
D] 0.02 mm

58] 1000 microns means -----
A] 1 mm
B] <u>1 m</u>
C] 1000 mm
D] 10 cm

59] in a metric micrometer, a complete revolution of
thimble advances -----------
A] 0.01 mm

B] 0.25 mm
C] 0.50 mm
D] 100mm

60] Ratchet Stop in the micrometer helps to ------------
A] Control the pressure
B] lock the spindle
C] Adjust the zero error
D] Hold the work piece

61] 1000 micron means ------------
A] 1 mm
B] 1 m
C] 1000 mm
D] 10 cm

62] What is the zero reading of a 50-75 mm outside micrometer?
A] 0000 mm
B] 001 mm
C] 2500 mm
D] 5000 mm

63] The value of the smallest division on sleeve of a metric outside micrometer is -----
A] 050 mm
B] 100 mm
C] 150 mm
D] 200 mm

64] Ratchet stop in the micrometer helps to ---------
A] control the pressure

B] Lock the spindle
C] Adjust the zero error
D] Hold the work piece

65] The least count of a vernier bevel protractor is
A] 1"
B] <u>5'</u>
C] 1°
D] 5 °

66] The part of a vernier bevel protractor which is normally used as a reference base for measuring angles is the
A] Blade
B] <u>Stock</u>
C] Disc
C] Main scale

67] The part of a vernier bevel protector on which main scale divisions are marked is the
A] Stock
B] Dial
C] <u>Disc</u>
D] Adjustable blade

68] The part of a bevel protractor, which comes in contact with the inclined surface while measuring is the
A] <u>Blade</u>
B] Stock
C] Disc
D] Dial

69] The value of each division of the main scale of a vernier bevel protractor is
A] 5'
B] <u>1°</u>
C] 5°
D]10°

70] The value of each division of the vernier scale of a bevel protractor is
A] 1°
B] 1°5'
C] <u>1°55'</u>
D] 5'

71] The taper shank drills are held on the machine by means of
A] Chucks
B] <u>Sleeves</u>
C] Drift
D] Vice

72] Drill chucks are fitted on the drilling machine spindle by means of a
A] Knurled ring
B] <u>Arbor</u>
C] Drift
D] Pinion and key

73] The Morse taper provided on drills ranges between
A] <u>MT 1 to MT 5</u>
B] MT 1 to MT 4
C] MT 0 to MT 5

D] MT 0 to MT 4

74] A drift is used for
A] Drawing a drill location
B] Fixing chuck on the machine spindle
C] Removing a broken drill from the work
D] <u>Removing the drill from the machine spindle</u>

75] When the taper shank of the drill is larger than the machine spindle, the device to hold the drill is a
A] Drill sleeve
B] <u>Taper socket</u>
C] Drill drift
D] Chuck and key

76] The suitable cutting fluid for drilling mild steel in a drilling machine is
A] Synthetic soluble oil
B] Neat oil
C] Distilled water
D] <u>Soluble oil</u>

77] A special feature of the radial drilling machine is
A] It can be used for drilling with a HSS drill
B] Table can be moved and set at any position
C] A variety of speeds is available
D] <u>The spindle can be brought to any position</u>

78] The point angle of drills depends on
A] The size of the drill
B] The type of machine
C] <u>The material of the work</u>

D] The RPM of the drill

79] The point angle for a standard drill is
A] 60°
B] 108°
C] <u>118°</u>
D] 135°

80] The helical angle determines the
A] Cutting angle
B] Chew angle
C] <u>Rake angle</u>
D] Lip angle

81] The clearance angle of the drill is between
A] 3° to 5°
B] <u>8° to 12°</u>
C] 12° to 20°
D] 15° to 20°

82] In a remote place (no electricity available] a rail track is to be drilled Choose the right drilling machine
A] Radial drilling machine
B] Pillar drilling machine
C] <u>Ratchet drilling machine</u>
D] Sensitive drilling Machine

83] A drilling machine used by a carpenter for cabinet making is a
A] Ratchet drilling machine
B] Radial drilling machine
C] <u>Breast drilling machine</u>

D] Sensitive drilling machine

84] Which one of the following drilling machines is used for drilling holes where electricity is not available?
A] Bench drilling machine
B] Pillar drilling machine
C] Redial drilling machine
D] Ratchet drilling machine

85] Which one of the following drilling machine is used for heavy duty work?
A] Bench drilling machine
B] Pillar drilling machine
C] Radial drilling machine
D] Electric hand drilling machine

86] Drill chuck are held on the machine spindle by means of ------
A] arbor
B] Drift
C] draw-in bar
D] Chuck nut

87] Different speeds are obtained in a sensitive bench drilling machine by ----
A] Belt pulley mechanism
B] Hydraulic mechanism
C] Rack and Pinion mechanism
D] Cam and follower mechanism

88] Tap are re sharpened by grinding -----
A] Hutes

B] Threads
C] Diameter
D] Relief

89] The tapping drill size for M10 x 15 is ----------
A] 82
B] 83
C] 84
<u>D] 85</u>

90] A nut is to be made for a screw of M10XIS What should be the size of drilled hole?
<u>A] 8-5 mm</u>
B] 90 mm
 C] 95 mm
D] 100 mm

91] The process of enlarging the end of a hole for accommodating the socket screw head is
A] Reaming
B] Spot facing
C] <u>Counter boring</u>

92] Appropriate tool used for spot facing operation is
A] Reamer
B] Counter sinks
C] <u>Fly cutters</u>

93] A short reamer with an axial hole used with an arbor or mandrel is called -------
A] Parallel reamer
B] Adjustable reamer

C] Expansion reamer
D] Chucking reamer

94] Which one of the following machine reamers is used to correct the misalignment between the reamer axis and the work axis?
A] Floating blade reamer
B] Machine jig reamer
C] Shell reamer
D] Chucking reamer

95] The depth of cut for metric square threading is
A] 06 x P
B] 05 x P
C] 05412 x P
D] 06412 x P

96] To cut buttress thread, the depth of cut is
A] 05412 x P
B] 06 x P
C] 07 x P
D] 075 x P

97] The Gear ratio required for cutting a screw thread of 25 mm on a lathe having a lead screw pitch using single point cutting tool is ----
A] 1:2
B] 2:1
C] 1:1 mm

98] A tumbler gear unit has
A] a single gear

B] two gears
C] <u>three gears</u>
D] four gears

99] Which one is the operation that cannot be done on the slotting machine?
A] key way slotting
B] dovetail slotting
C] gear cutting
D] <u>thread cutting</u>

100] The threads on the back side of the four Jaw chuck has type-----of threads
A] <u>Square</u>
3] Trapezoidal
C] V -shape
D] None of these

101] A voltage source produces an IR drop of 40V across a 20 ohms resistance, 60V across a 30 ohms resistance and 180V across a 90 ohms resistance all in series]How much is the applied voltage?
A] 180 V
B] 240 V
C] 100 V
D] <u>280 V</u>

102] The initial function of a choke in a tube light circuit is to…
A] limit the starting current
B] <u>induce high voltage</u>
C] heat up the filament

D] limit the current after starting

103] The peak-to-peak voltage is 99V]how big is the effective value of the sine wave?
A] 70 V
B] 44.5V
C] 49.5 V
D] <u>35 V</u>

104] A moving coil voltmeter reads 10 V AC]How big is the effective voltage?
A] higher
B] lower
C] <u>the same</u>
D]10% higher

105] A capacitor is connected across a 200 volt AC line, its minimum voltage rating should be...
A]100 volts
B] 200 Volts
C]<u>300 volts</u>
D]400 volts

106] How much is the nominal output voltage of a carbon zinc cell?
A] 12V
B] <u>1.5V</u>
C] 2.0V
D] 2.2V

107] Cells are connected in series to..
A] <u>increase the output voltage</u>

B] decreases the output voltage
C] decrease the internal resistance
D] increase the current capacity

108] An unknown DC voltage is to be measured, which measuring range will you select first?
A] 500V
B] 50V
C] 1.5 V
D] 0.5V

109] Heat developed in a conductor is proportional to the…
A] square of the power
B] square of the resistance
C] square of the current
D] square of the time

110] The second function of a choke in a tube light circuit is to…
A] limit the starting current
B] induce high voltage
C] heat up the filament
D] limit the current after starting

111] A moving iron ammeter reads 10 A]how big is the peak current of the oscillation?
A] 7.07 A
B] 1.1414A
C] 70.7 A
D] 14.1 A

112] Power companies are interested in improving the power factor to
A] <u>reduce line current</u>
B] increase motor efficiency
C] increase volt-amperes
D] decrease power

113] In a RL parallel circuit, the opposition to total current is called…
A] reactance
B] resistance
C] a vector sum
D] <u>impedance</u>

114] An unknown direct current of micro ampere rating is to be measured, which measuring range will you select first?
A] 20 micro amp
B] 15 micro amp
C] 150 micro amp
D] <u>500 micro amp</u>

115] The earth conductor provides a path to ground for..
A] <u>leakage current</u>
B] over current
C] high voltage
D] circuit current

116] Which appliance works on heating effect of electric current?
A] incandescent lamp
B] bimetallic thermostat

C] H R C fuse

D] <u>toaster</u>

117] External Thread provide on Rod or Pipe , by Die and Cutting Tool is called ......

(A] Tapping

(B] Dieing

(C] <u>Threading</u>

(D] Grooving

118] G.l pipes are provided externally with

A] no threads

B] <u>parallel threads</u>

C] tapered threads

D] neither parallel nor tapered threads.

119] in the pipe assembly, the hemp packing is used

A] for easy engagement

B] to fill the gap between threads

C] <u>to avoid leakage</u>

D] to get tight fitting.

120] The sealing compound shall be applied on the pipe threads

A] before hemp packing

B] <u>after hemp packing</u>

C] before and after temp packing

D] none of the above.

121] Used on finished tubular wrench surfaces to avoid marking]

A Stillson pipe

B] <u>Chain wrench</u>
C] Strap wrench
D] Footprint wrench

122] Used for gripping and turning pipes and round stocks in confined places]
A] Stillson pipe
B] Chain wrench
C] Strap wrench
D] <u>Footprint wrench</u>

123] Used for holding large diameter pipes]
A] Stillson pipe
B] <u>Chain wrench</u>
C] Strap wrench
D] Footprint wrench

124] Used for gripping and turning pipes,tubes and cylindricai rods]
A] <u>Stillson pipe</u>
B] Chain wrench
C] Strap wrench
D] Footprint wrench

125] Secures rope to small pipe or rim.
A] Slip knot
B] Bowline knot
C] Square knot
D] <u>Sheep shank knot</u>]

126] It can be folded and carried to any place] Similar to the quick releasing type pipe vice.

A Portable folding pipe vice
B] Chain pipe vice
C] Pipe vice
D] None of above

127] Used to hold pipes more than 63mm to 200mm diameter.
A] Portable folding pipe vice
B] Chain pipe vice
C] Pipe vice
D] None of above

128] Used for quick holding and locating pipes] Used to hold pipes up to 63mm diameter]
A] Portable folding pipe vice
B] Chain pipe vice
C] Pipe vice
D] None of above

129] Provides deviation of 90°
A] Plug
B] Elbow
C] Bend
D] Reducer 'T' branczh

130] Provides change of direction with a long radius at right angle.
A] Plug
B] Elbow
C] Bend
D] Reducer 'T' branczh

131] Used for closing a line which has an internal thread.
A] Plug
B] Elbow
C] Bend
D] Reducer 'T' branczh

132] Provides deviation of 45°
A] Bend
B] Reducer 'T' branczh
C] Elbow
D] Tee piece

133] Provides outlet at right angles to the run.
A] Bend
B] Reducer 'T' branczh
C] Elbow
D] Tee piece

134] Used where a change in ' pipe diameter is required]
A] Bend
B] Reducer 'T' branczh
C] Elbow
D] Tee piece

135] Selection of a former depends on the
A] outside diameter of the pipe
B] wall thickness of the pipe
C] bore diameter of the pipe
D] all the above.

136] A branch type hand operated pipe bending machine is
used to bend

A] P.V.C.pipes
B] onduit pipes
C] <u>G.I.pipes</u>
D] copper pipes.

137] The inner formers of a hydraulic pipe bending machine are able to bend pipes up to a diameter of
A] 40mm
B] 100mm
C] 20mm
D] <u>75mm</u>

138] The included angle of a pipe thread is
A] 60°
B] 47°
C] <u>55°</u>
D] 45°

139] G.l.pipes are available in a standard length of
A] 5 metres
B] 18"
C] <u>6 metres</u>
D] 16 feet.

140] The standard pipe fittings are provided with threads conforming with
A] BA
B] BSW
C] <u>BSP</u>
D] Metric.

141] The external threads on G.l.pipes are out easily

A] by tap sets
B] dies and die stocks
C] centre lathes
D] thread rollers

142] Used where bolt and threads are to be protected from damage.
A] Donald cap nut
B] Thumb nut
C] Hexagonal nut
D] Wing-nut

143] Used where frequent removal and fixing is required.
A] Donald cap nut
B] Thumb nut
C] Hexagonal nut
D] Wing-nut

144] Used in machine building and structure work.
A] Donald cap nut
B] Thumb nut
C] Hexagonal nut
D] Wing-nut

145] Used where frequent adjustments are to be made.
A] Donald cap nut
B] Thumb nut
C] Hexagonal nut
D] Wing-nut

146] Nylon inserts in the nut prevent loosening.
A] Locking plate

B] Wire lock
C] <u>Self-locking nut</u>
D] Sawn nut

147] A slot is cut halfway across the nut.
A] Locking plate
B] Wire lock
C] Self-locking nut
D] <u>Sawn nut</u>

148] Prevents slackening of two bolts.
A] Locking plate
B] <u>Wire lock</u>
C] Self-locking nut
D] Sawn nut

149] Prevents rotation of the top nut.
A] <u>Lock-nut</u>
B] Grooved nut
C] Self-locking nut
D] Sawn nut

150] Prevents loosening of nut by the use of a plate shaped to fit the nut.
A] <u>Locking plate</u>
B] Wire lock
C] Self-locking nut
D] Sawn nut

151] Hexagonal nut with the lower part made cylindrical and the recessed groove.
A] Lock-nut

B] <u>Grooved nut</u>
C] Self-locking nut
D] Sawn nut

152] Drill a blind hole equal to half of the diameter of the stud. Insert this tool into the hole and remove the stud by turning this anticlockwise.
A] Prick Punch Method
B] Filing square very mm
C] <u>Using square taper punch</u>
D] Ezy-out method

153] If the stud is broken near to the surface, employ this method to remove the stud.
A] <u>Prick Punch Method</u>
B] Filing square very mm
C] Using square taper punch
D] Ezy-out method

154] When a stud is broken a little above the surface this method is used to remove the stud.
A] Filing square very mm
B] Using square taper punch
C] Ezy-out method
D] <u>Making drill hole</u>

155] To extract the broken stud a special tool is employed in this method.
A] Prick Punch Method
B] Filing square very mm
C] Using square taper punch
D] <u>Ezy-out method</u>

156] The inner formers of a hydraulic pipe bending machine are able to bend pipes up to a diameter of
A] 40mm
B] 100mm
C] 20mm
D] <u>75mm</u>

157] Which is not the property of hydraulic fluid used in grinding machine?
A] it must not control or absorb air
B] it must not cause corrosion of the moving parts
C] Should have adequate viscosity
D] <u>it must vaporize at the operating temperature</u>

158] Which one of the following is the advantage of pneumatic system?
A] <u>For low cost layout</u>
B] For increasing the rate of production
C] For better working environment

159] Following which advantage of Pneumatic power system
A] For increase production rate.
B] Less cash for layout
C] Good climate for work
D] <u>Above all</u>

160] The pressure of fluid in hydraulic brake system is governed by
A] boils law
B] Charles law

C] <u>Pascal's law</u>
D] none of the above laws

161] Allows fluid both way in and out of cylinder
A] Piston
B] Push Rod
C] Primary cup
D] <u>Check valve</u>

162] Relieves excess pressure of air from the air tank]
A] Air compressor
B] Unloader valve
C] <u>Safety valve</u>
D] Brake chamber

163] Regulates maximum air pressure, reaching to air tank]
A] Air compressor
B] <u>Unloader valve</u>
C] Safety valve
D] Brake chamber

164] Distributes air to various circuits
A] Brake actuator
B] Dual brake valve
C] <u>System protection valve</u>

165] Develops pressure on fuel to go out
A] Valves
B] Coil spring
C] <u>Diaphragm</u>
D] Rocker arm

166] Takes thrust load
A] Crankshaft
B] Flywheels
C] Torque wrench
D] <u>Thrust bearing</u>

167] Only spur gears are used
A] <u>Sliding mesh</u>
B] Synchromesh
C] Double declutching
D] Transfer case

168] Used for smooth gear shifting
A] Sliding mesh
B] Synchromesh
C] <u>Double declutching</u>
D] Transfer case

169] Hard gear shifting is due to
A] Worn out clutch disc
B] Damaged main shaft bearings
C] <u>Synchronizer unit damaged</u>
D] excessive oil in the gearbox.

170] Gear slip is due to
A] <u>Worn out synchroniser</u>
B] Worn out clutch disc
C] Dry main shaft bearing
D] Weak pressure spring of clutch.

171] Noise in particular gear is due to
A] Insufficient clutch pedal free play

B] Damage gear teeth

C] Cracked gear box case

D] <u>Damaged synchromesh unit</u>.

40